BEGINNINGS

THE EARTH

ORIGINS AND EVOLUTION

Evolution of the Universe

4.5 billion years ago the oceans and first landmasses form. **9**

1 million years after the Big Bang, hydrogen atoms form. **5**

10-20 billion years ago, in less than a second, four things happen.

1. The Big Bang
2. Inflation
3. The beginning of the four forces
4. The first atomic nuclei form

1 billion years after the Big Bang, galaxies begin to form. **6**

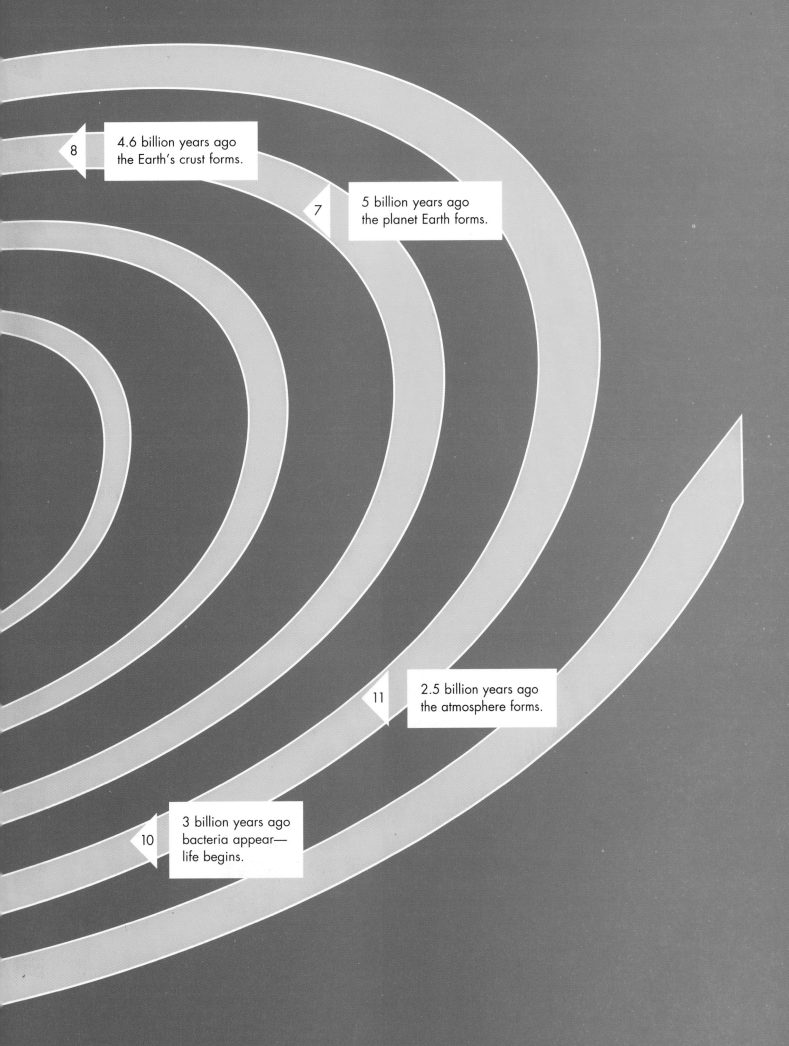

BEGINNINGS

THE EARTH

ORIGINS AND EVOLUTION

by
Anna Alessandrello

English Translation by Rocco Serini

RSVP
**RAINTREE
STECK-VAUGHN**
P U B L I S H E R S
The Steck-Vaughn Company

Austin, Texas

Published by Raintree Steck-Vaughn Publishers, an imprint of Steck-Vaughn Company

Series Editor: Caterina Longanesi
American Edition, Edit and Rewrite: Susan Wilson
Consultant: Thomas A. Davies, Deputy Director, Institute for Geophysics, University of Texas
Project Manager: Julie Klaus
Electronic Production: Scott Melcer
Cover Artwork: Antonio Molino

Photographs ALESSANDRO BOESI, Milan: p. 30 (1, 2), p. 32 (2). ANDREA BORGIA, Milton Keynes, UK: p. 12 (1, 2). FRANCO CAPONE, Milan: p. 33 (3). ALBERTO CONTRI, Milan: p. 17 (3), p. 23 (2, 3), p. 37 (3), p. 33 (5), p. 41 (4, 5). Editoriale Jaca Book, Milan (Carlo Scotti): p. 11 (2), p. 14 (2, 3, 4, 5), p. 15 (6, 7, 8, 9, 10, 11, 12, 13, 14, 15). NASA, Washington, DC: p. 10 (1), p. 35 (2), p. 37 (2). GIORGIO NOSOTTI, Milan: p. 29 (4), p. 23 (4). STEFANIA NOSOTTI, Milan: p. 29 (5). GIANNI PEROTTI, Milan: p. 14 (1), p. 17 (2). GIOVANNI PINNA, Milan: p. 11 (3), p. 27 (2), p. 41 (3). MADDALENA POCCIANTI SCANDICCI, Florence: p. 28 (1, 2, 3), p. 27 (3). CARLO SCOTTI, Milan: p. 37 (4). GEORGIO TERUZZI, Milan: p. 35 (3).

Illustrations: Editoriale Jaca Book, Milan (Cesare Dattena): p. 40 (1, 2), stages, p. 46, p. 47; (Maria Elena Gonano): p. 29 (6), p. 32 (1), p. 33 (4); (Maurizio Gradin and Fabio Jacomelli): p. 11 (1); (Antonio Molino): p. 24; (Rosalba Moriggia and Maria Piatto): p. ii, p. iii, p. 8, p. 12 (stages), p. 16 (1), p. 18. p. 36 (1), p. 38, p. 20, p. 26, (1), p. 31 (3); (Lorenzo Orlandi): p. 13 (3), p. 34 (1).
Illustration p. 22 (1) furnished by Cherrytree Press LTD, Bath, Avon, UK, who published it in *Oceans* by Tom Mariner, illustrated by Mike Atkinson, 1989.

Graphics and Layout: The Graphics Department of Jaca Book
Special thanks to the Museum of Natural History of Milan

Library of Congress Cataloging-in-Publication Data
Alessandrello, Anna.
    [Terra. English]
    Earth: origins and evolution / by Anna Alessandrello.
       p.   cm. — (Beginnings)
    Translation of: La Terra.
    Includes index.
    ISBN 0-8114-3331-5
    1. Earth — Origin. 2. Geology. 3. Geophysics. I. Title. II. Series. Beginnings (Austin, Tex.)
QB632.A3813   1995
550—dc20   94-2874
                                                  CIP
                                                  AC

Printed and bound in the United States

1 2 3 4 5 6 7 8 9 0 KP 99 98 97 96 95 94

TABLE OF CONTENTS

THE EARTH IS BORN

If you look around right now, you might see buildings, trees, rocks, or things that make up the Earth as we know it. But what was the Earth like before buildings, before trees, and even before rocks? A long time ago, billions of years in the past, the Earth was merely a mass of swirling dust and gas.

The Earth revolves around the sun at a distance of about 93.7 million miles, and its radius measures about 4,000 miles. Scientists believe that the Earth, and the other eight planets of the solar system, formed as the sun itself formed. This theory is best explained by the **nebular hypothesis** of the origin of the solar system. At first a huge cloud of gases and cosmic dust slowly moved through the universe. Then about 5 billion years ago, the gravitational force of the cloud caused it to begin to contract. As the cloud contracted, it began to spin. As the mass of dust and gas continued to condense, it spun faster and faster, flattening the cloud into the shape of a disk. At the same time, gravity caused most of the matter to collect in a large mass at the center of the cloud, a large mass which would become the sun. This large mass grew hotter and hotter, forming a star we call the sun.

Within the outer areas of the disk, small whirlpools formed where matter also collected. These whirlpools of matter that orbited the sun became **protoplanets**, swirling masses of dust and gas.

All the protoplanets within the cloud were probably made up of the same elements, or basic chemical materials. But the four protoplanets closest to the sun were affected by powerful solar

5 billion years ago the Earth was formed as part of the solar system.

4.6 billion years ago the Earth's crust formed.

4.5 billion years ago the ocean and original landmasses formed.

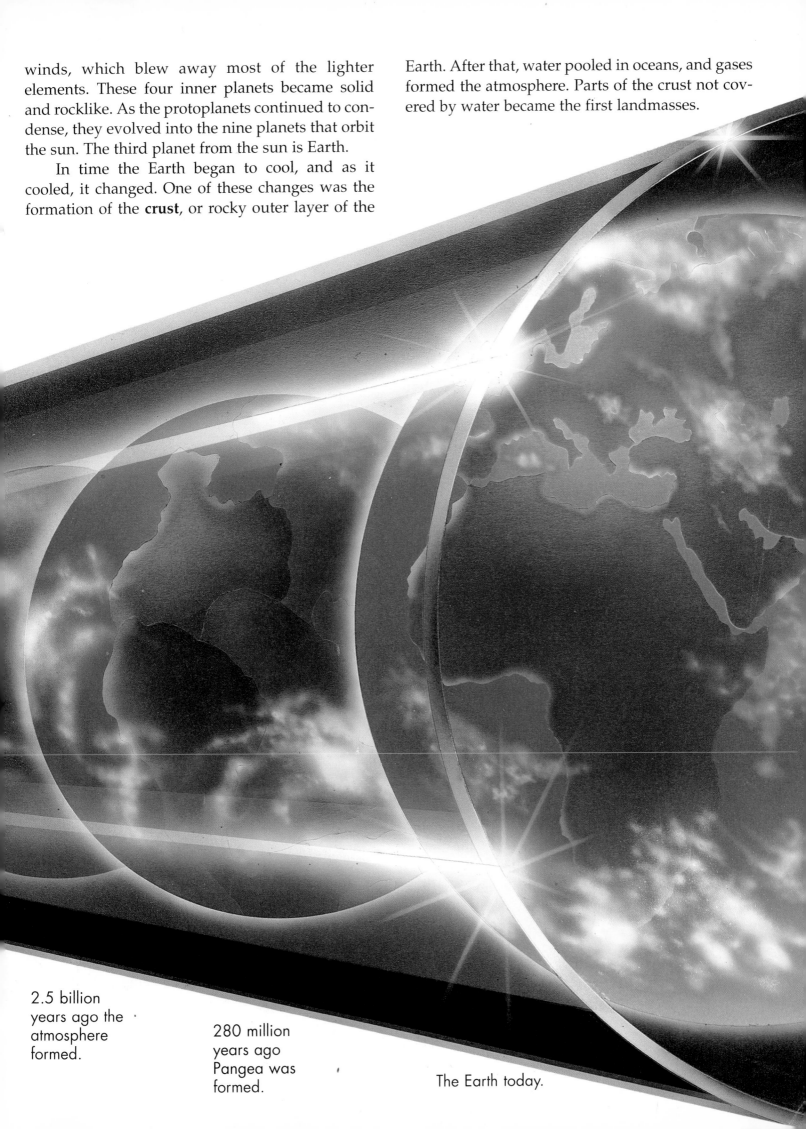

winds, which blew away most of the lighter elements. These four inner planets became solid and rocklike. As the protoplanets continued to condense, they evolved into the nine planets that orbit the sun. The third planet from the sun is Earth.

In time the Earth began to cool, and as it cooled, it changed. One of these changes was the formation of the **crust**, or rocky outer layer of the Earth. After that, water pooled in oceans, and gases formed the atmosphere. Parts of the crust not covered by water became the first landmasses.

2.5 billion years ago the atmosphere formed.

280 million years ago Pangea was formed.

The Earth today.

STRUCTURE OF THE EARTH

In photos taken from space, the Earth looks like a dark, blue and brown ball shrouded in clouds. Beneath the clouds and the transparent atmosphere lies the Earth's crust, the outermost, solid layer of the Earth. The crust is a very thin, rocky layer that forms a brittle skin around the inner layers of the Earth. The thickest parts of the crust are found in large mountain ranges where the crust is about 20 to 25 miles thick. But the Earth's crust also extends beneath the ocean. The crust may be only about 3 miles thick under the oceans.

Below the crust is a thicker, rocky layer called the **mantle**, which is about 1,800 miles thick. Parts of the mantle are made up of molten, or melted, rock. Because the molten rock is a liquid, it flows. This liquid layer of rock can cause movement and breaks within the brittle crust above it. The movement of the mantle is the cause of the evolution of the Earth's landmasses, as well as the cause of earthquakes and volcanoes. You will read more about these subjects in the next chapter.

The mantle surrounds a layer called the **outer core**, which is about 1,400 miles thick. Scientists have little direct information about this layer but have found clues to its makeup in meteorites. Meteorites are objects that fall to the Earth's surface from space and are believed to be made up of the same materials which form the Earth. Some meteorites that have a large concentration of iron and nickel. Because these elements are found in much smaller concentrations in the upper layers of the Earth, scientists believe that they are probably very concentrated in the **core**.

Within the outer core is the **inner core**, the innermost structure of the Earth. Like the outer core, the inner core is probably made up of a mixture of iron and nickel. Unlike the outer core, the inner core is believed to be solid, not molten.

1. The Earth is made up of a series of layers which are surrounded by the Earth's atmosphere, a layer of gases held by the Earth's gravity. The layers of the Earth are the crust, mantle, outer core, and inner core.

1

inner core

outer core

mantle

crust

2. Granite, a rock formed from materials in the mantle, is commonly found in the Earth's crust. Different elements make up granite and can be seen in the different-colored crystals.
3. Close-up of a meteorite showing its iron composition.

3

The Dynamic Earth

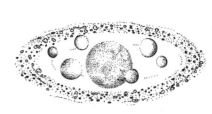

5 billion years ago the Earth was formed as part of the solar system.

4.6 billion years ago the Earth's crust formed.

When the Earth first formed, it may have looked like a small version of the sun—a ball of fire in space. This blazing protoplanet was made up of many different elements, or basic chemicals. Because the heat was so intense, many of these elements were in a molten, or liquid, form, even those that you usually think of as solids, such as copper or silver. During this time, the heavier elements, such as iron and nickel, sank toward the center of the Earth, forming the inner and outer cores. These heavy metals were the first to solidify. The lighter elements floated up to the outer layers, forming large parts of the mantle and crust. Still lighter elements, such as hydrogen and oxygen, were in gas form and bubbled up through the layers. Some of this gas escaped from the Earth. It was too light to be held by the Earth's gravity and was blown away by a wind radiating from the sun.

Elements that were too light to be pulled to the core and were too heavy to drift into space formed new compounds. Elements, such as **silicon**, magnesium, potassium, and sodium, combined to form **minerals**. Minerals in turn formed rocks that make up the crust and the mantle. Some rock within the upper layer of the mantle is in molten form and is called **magma**. When enough pressure builds up within the mantle, magma flows through cracks in the Earth's crust, resulting in a volcano. Volcanoes and **geysers** provide glimpses of the Earth as it was when it was forming, with materials flowing from one layer to another. The Earth is a dynamic system that continues to evolve. Volcanoes are one part of this evolutionary process.

1

1. Mt. Etna, a volcano in Sicily, Italy. The lava, or molten rock, flowing from the volcano, moves at a speed of about 8 inches per second and has a temperature of about 1,800° F.
2. Because lava solidifies as it cools, the upper layer hardens first. Here hot, molten lava flows through cracks in the hardened, upper surface.

2

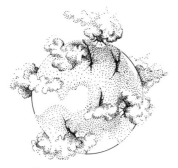

4.5 billion years ago the ocean
and original landmasses formed.

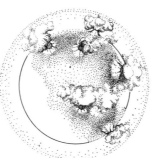

2.5 billion years ago
the atmosphere formed.

280 million years ago
Pangea was formed.

3. The newly-evolving Earth rotated and revolved around the
sun as it slowly began to cool. Various chemicals gradually
formed the minerals and rocks of the Earth's crust. The crust
was probably very thin, having wide cracks and deep
openings. From these openings the burning lava and gases
flowed. Impressive canals of lava running down the sides of
an erupting volcano give us an idea of what the Earth may
have looked like 4.6 billion years ago.

ROCKS

Because the Earth's crust is largely made up of rocks, the study of rocks is a basic part of any study of the Earth. All rocks are made up of one or more types of minerals. A mineral is a solid element or compound that has its own specific crystal structure. Some minerals, such as a diamond, are made up of a single element. Other minerals, such as table salt, are compounds. In the same way, a rock

As igneous rock is broken down by wind, water, glaciers, plants, or other forces, it forms **sediment**, or small bits of rock and dust. In time, sediment may be blown or washed away. Much of it collects at the bottom of the ocean. As the layers of sediment are buried by more sediment, pressure and chemicals from the water cement the sediment particles to form **sedimentary rock**, the second type of rock.

If sedimentary rock gets buried deep within the Earth and is under greater pressure as well as

6

8

sedimentary rock

igneous rock

metamorphic rock

2   3

1 4   5

10

may be made up of a single mineral or a mixture of minerals. **Granite**, with its many specks of color, is an example of a rock made up of a mixture of minerals. Each different-colored speck is the crystalline form of a different mineral.

Rocks are classified by the way they form. The first type, **igneous rock**, forms as magma cools. This is the type of rock that scientists think made up the Earth's crust when it first evolved. The molten material cooled when it came to the surface.

under great heat, it will turn into the third type of rock, **metamorphic rock**.

If metamorphic rock moves still deeper into the Earth, it would be under greater pressure and heat and would melt to form magma. If this magma should move to the Earth's surface, it would once again cool and change to igneous rock. This cycle of magma to solid rock and back again is called the **rock cycle**.

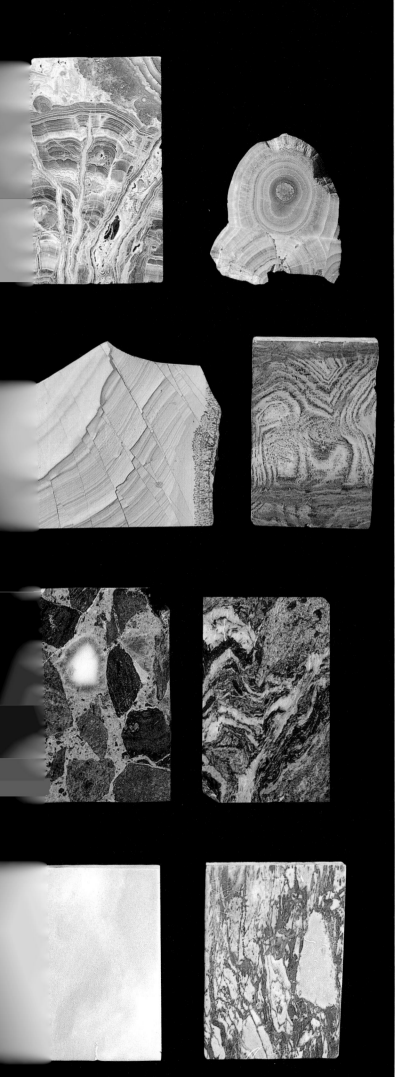

Rocks are sometimes found in strange shapes and formations, such as this rock (1) in Arches National Park, Utah. Weathering has worn away much of a larger rock, leaving behind this strangely-shaped rock.

Porphyry (2), gabbro (3), and granite (4 and 5) are all types of igneous rock.

7

14

9

15

11

From these came sedimentary rocks like alabaster (6 and 7) and limestone (8 and 9). Metamorphic rocks shown here are serpentinite breccia (10), serpentinite (11), and marble (12 and 13). Calcite (14) and quartz (15) are both minerals. Notice that minerals form in the shape of crystals.

13

LAYERS OF HISTORY

Sometimes the history of the Earth's formation is laid open for us like the pages of an enormous book. This can happen when a river cuts a steep-sided valley through a mass of rock, or when earth movements crack a mountain down the middle. Then we may see **strata**, or broad layers of sedimentary rock. These strata were laid down, one on top of the other, as the Earth evolved. Normally we only see the top of the uppermost layer—the one most recently laid down. Now, when we are able to see the strata from the side, it is as if we are seeing a slice of time.

Examining these strata can help us reconstruct a picture of the early Earth and see how it changed over time. We begin to read this story from the bottom, where the oldest layers formed, and gradually read upward toward the youngest layers. Sometimes one layer of rock may hold fossil remains of sea animals, such as **mollusk** shells or fish bones. That would tell us that this area was once covered by water, possibly the ocean. A higher layer may hold coal and traces of plant life. This would show not only that this area was later land, but would also tell about the type of plants living then. From this evidence we could determine changes in the climate. Higher strata might show evidence of other changes in climate, such as layers made up of pebbles from a river or sand from a desert.

1. We can reconstruct the history of a mountain by studying its strata. This is an example of strata that could be built up over time, with the oldest layers from the Ordovician Period being laid down first, and those from the Triassic, the youngest period, being laid down last. Notice that some strata may include more than a single type of rock, showing folds or fault lines in the rock.
2. Weathering, or the wearing away of the land, can be caused by wind or by water. This exposes some of the rock that makes up the Earth's crust. Various strata are clearly visible in Arches National Park, Utah.
3. A mountain chain has been so greatly weathered and eroded that only small parts still remain in Monument Valley, Arizona.

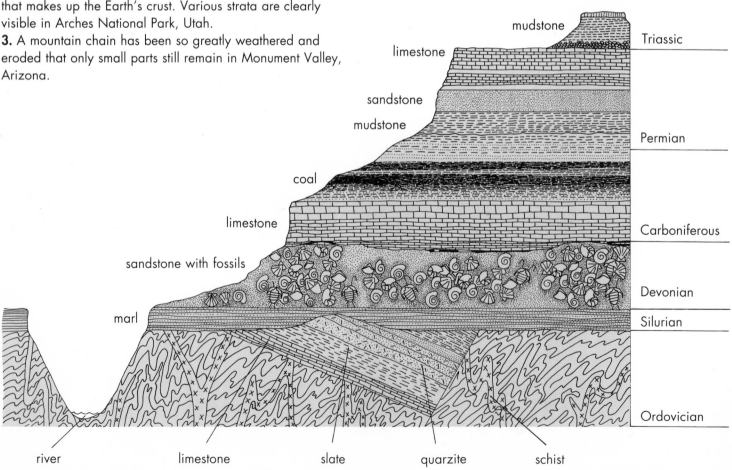

16

Birth of Land and Ocean

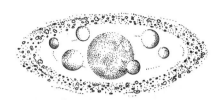

5 billion years ago the Earth was formed as part of the solar system.

4.6 billion years ago the Earth's crust formed.

When the Earth first formed from a mass of gas and dust, many different elements came together. These included hydrogen and oxygen, the two elements that make up water. As the burning proto-planet cooled, the heavier elements moved to the Earth's core, while hydrogen and oxygen and other light elements moved outward. As the Earth cooled, elements joined together to form compounds. At this time water was probably trapped within the magma of the young Earth.

As the Earth continued to cool, magma rose from the mantle and solidified into igneous rocks, forming the Earth's crust. Deep cracks formed in the crust as a result of uneven cooling. Lava, or magma that has reached the surface, spewed out of the cracks and erupted from volcanoes. The surface of the Earth was constantly changing. As the lava cooled, rocks and minerals would have solidified while water escaped as a gas—**water vapor**. Gravity held this gas within the Earth's early atmosphere. Once water vapor was in the atmosphere, it could rise up and cool. When water vapor cools, it condenses, or turns from a gas to a liquid. In this way, water vapor could be released from cooling rock as a gas and fall back to Earth as rain or another form of **precipitation**. Liquid water then pooled on the Earth's rocky crust, forming larger and larger bodies of water.

So much water fell back down to Earth, that most of the Earth's surface was covered. The rocky areas that remained exposed became the early continents. You will read about the movements of these early continents in the next chapter.

R. MORIGGIA & H. PIATTO

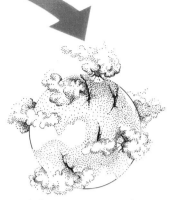

4.5 billion years ago the ocean and original landmasses formed.

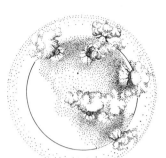

2.5 billion years ago the atmosphere was formed.

280 million years ago Pangea was formed.

THE MOVEMENT OF CONTINENTS

As the crust continued to evolve, oceans filled with water and large landmasses, or continents, rose above the oceans. Like the crust itself, the continents also continue to evolve. Scientists believe that there were several continents formed during the Earth's early history. Then about 280 million years ago these continents joined together to form one very large continent called Pangea.

Why do scientists think that these landmasses were once connected? For one thing, the shapes of some continents look as if they might fit together like pieces of a jigsaw puzzle. If you look at a world map, you will see that the coastlines of South America and Africa seem to have interlocking shapes. In addition, some mountain ranges and other land features seem to extend from one continent to another. A third reason to believe that today's separate continents were once linked is the similarity of animals on widely separated continents. These animals are too closely linked not to be related, and the landmasses are too far apart for the animals to have crossed the ocean. The formation and breakup of the supercontinent Pangea was explained in a theory called continental drift.

The theory of continental drift did not explain the forces causing continents to move. This movement is explained in a new theory called plate tectonics. This theory holds that the Earth's crust is composed of several **plates**. Each of these plates moves independently of the others. The plates move because of the movement of magma, the layer of hot, molten rock in the upper mantle. This movement causes continents to move, volcanoes and mountains to form, and the ocean floor to spread.

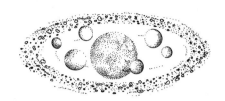

5 billion years ago the Earth was formed as part of the solar system.

4.6 billion years ago the Earth's crust formed.

1. It is thought that the Earth's landmasses first formed as separate continents. This shows the Earth as it may have appeared about 570 million years ago.
2. About 280 million years ago these landmasses are thought to have joined into one supercontinent called Pangea.

1

20

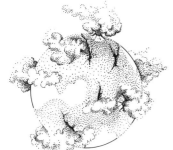
4.5 billion years ago the ocean and original landmasses formed.

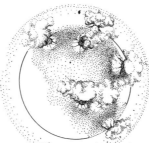
2.5 billion years ago the atmosphere was formed.

280 million years ago Pangea was formed.

2

THE OCEAN

The oceans of the world make up one giant body of water that covers over 70 percent of the Earth's surface. It is often referred to as various different seas and oceans. If you look at a world map or a globe, you will notice that all these major bodies of water connect. The water within all parts of the world's ocean is fairly uniform in composition. Ocean water is salt water—it is made up of about 3.5 percent salt. Most of that salt is sodium chloride, or table salt.

Below the surface of the water, the ocean floor is full of variety. Scattered along the bottom of the ocean are underwater mountains, volcanoes, **mid-ocean ridges**, and deep-ocean trenches. Just as similar structures form on land, these structures are the result of processes in the Earth's crust and the mantle.

As you read in the last chapter, the crust that surrounds the Earth is not a single, continuous piece, but it is broken up into many large pieces called plates. You can compare the Earth's crust to the shell of a hard-boiled egg after the shell has been cracked but before the egg has been peeled. Each separate piece of shell is still rigid and is still attached to the egg, but the pieces are not attached to each other. In the same way, the plates of the Earth's crust are rigid and are not attached to one another.

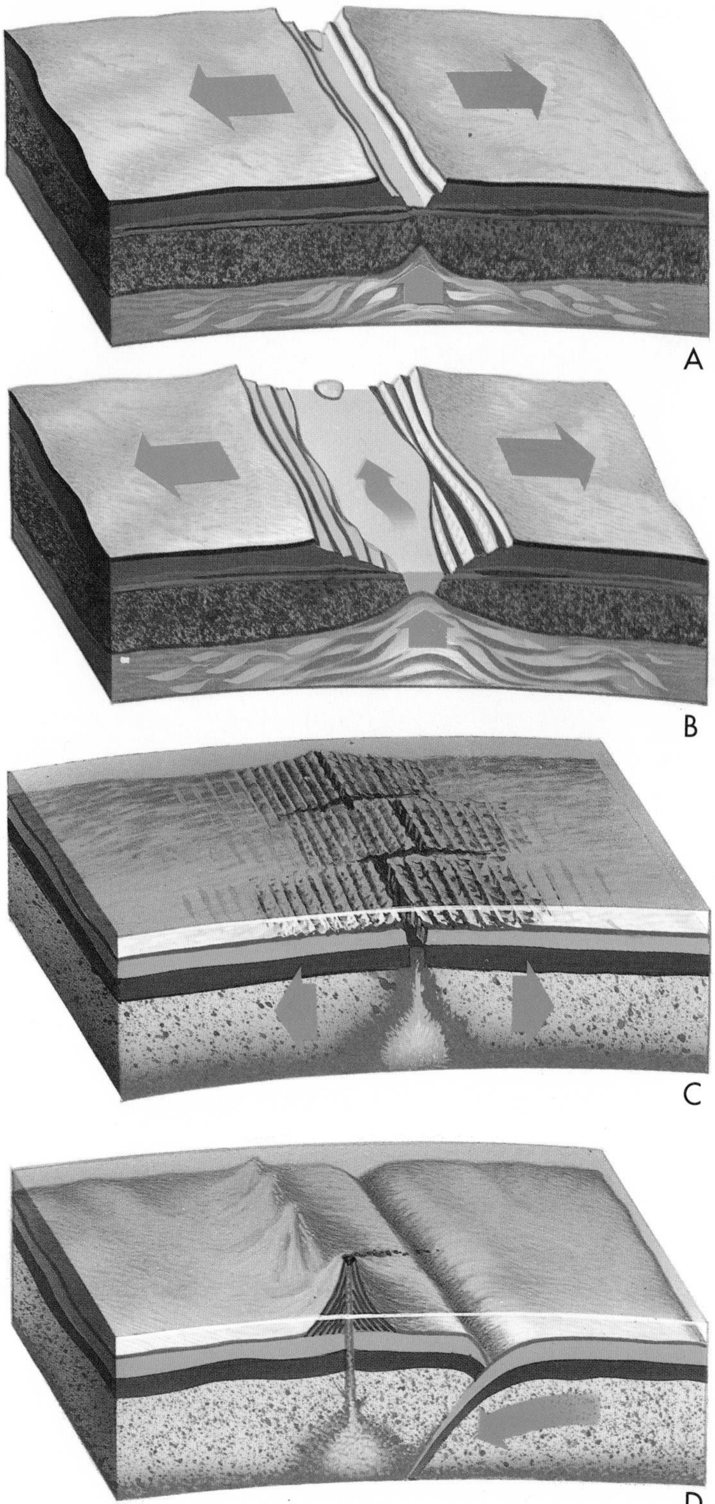

A

B

C

D

The diagram on pages 24-25 shows magma rising from a mid-ocean ridge, forming new ocean floor. This expanding plate pushes against a neighboring, continental plate (seen at the left). A deep-ocean trench forms where the oceanic plate sinks under the edge of the continental plate. As a result of new material being forced down into the mantle, volcanic islands form.

1. Crustal plates can move apart, as shown above in Figure A. As the plates continue to move apart, water flows in, forming a narrow sea, as shown in Figure B. The plates may continue to move apart, forming a broad ocean. The area where the new ocean floor continues to form is called a mid-ocean ridge, shown in Figure C.

In addition to moving apart, plates can move toward each other, as shown in Figure D. When this happens, one plate shifts below the other. The point where the plate sinks down is called a deep-ocean trench. Volcanoes form near the rift, caused by a plate sinking into the mantle.
2. A black beach, Stromboli, Italy. **3.** Cliffs in Stromboli.
4. A French Polynesian atoll.

The formation of a deep-ocean trench is shown at the left and a mid-ocean ridge at the right.

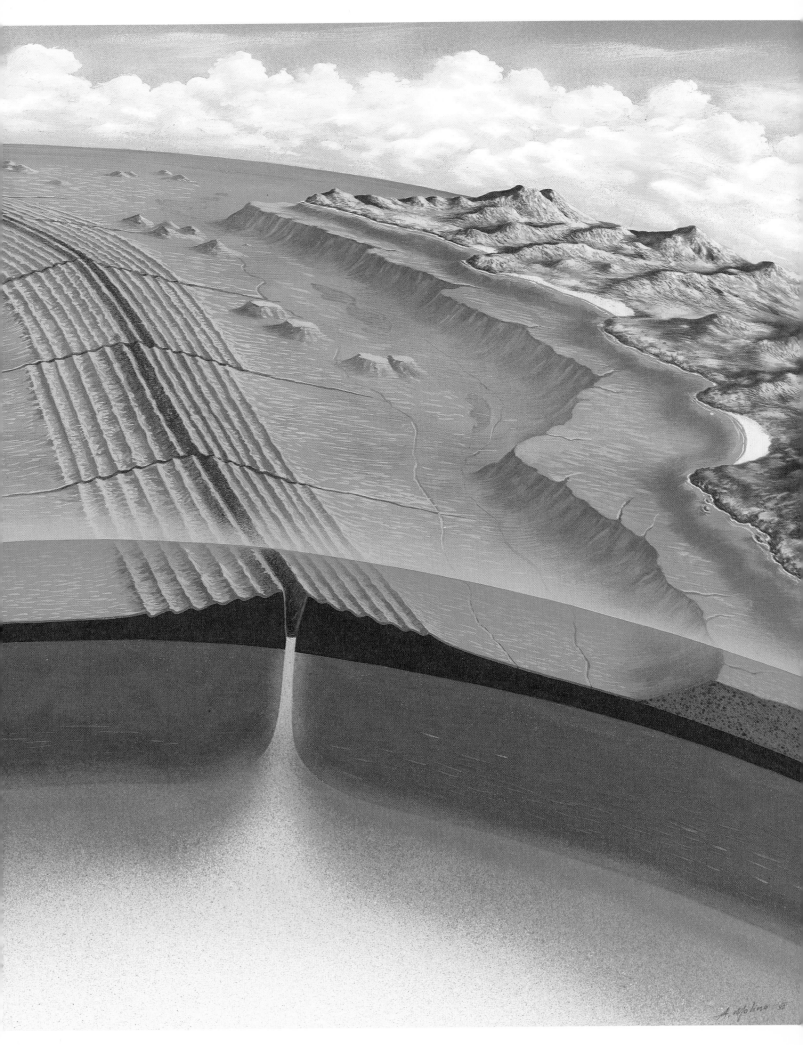

MOUNTAIN BUILDING

Large mountain ranges, with peaks rising as much as 5 miles above sea level, form when crustal plates collide. When two plates come together, their edges can crumple. Then the originally flat layers of rock making up the crust can bend and fold, like a tablecloth being pushed across a table. Under intense heat and pressure, the formerly rigid rock can change shape. Mountain ranges that form in this way are called complex or **folded mountains**. The major mountain ranges, including the Appalachians, Alps, and Himalayas, are all folded mountains.

Mountains may also form along **fault** lines, or breaks within a plate. These faults can form when crustal plates move away from each other. Later large chunks of rock can slip down, and other chunks can be pushed up. The Wasatch Range, in Utah, and the Teton Range, in the Rocky Mountains, are examples of **fault-block mountains**.

Mountains can also form as the result of volcanoes. When a volcano erupts, lava flows out, hardens, and forms a cone. Over time a large **volcanic mountain** can form.

Dome mountains, like the Black Hills of South Dakota, form when the Earth's crust is forced upward. These mountains form roughly circular bulges in the landscape. Similar structures that sink into the Earth are called basins.

Mountains can also form as the result of weathering, or a wearing away of the land. When rivers or glaciers cut through sedimentary rock, they can leave mountains and valleys in their path.

1. Folded mountains are also called complex mountains and form as a result of folding as well as faulting and metamorphic rock formation. Notice that the folds may be irregular and folded in on top of each other. Labels at the right show the time periods during which each layer of rock was laid down. The insert shows a fault-block mountain.

2. Anticlines and synclines are visible in this mountain. An anticline is a fold in the layer of rock that slants down on both sides. A syncline is a trough in the rock layers.
3. A fault and subsequent movement can be seen in this photo.

overlying layers

1

2

3

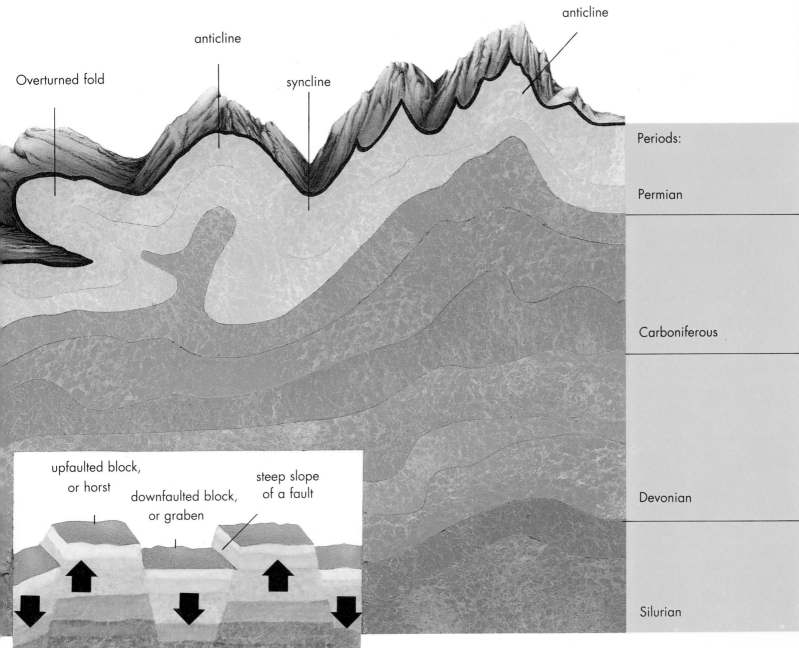

Overturned fold

anticline

syncline

anticline

upfaulted block,
or horst

downfaulted block,
or graben

steep slope
of a fault

Periods:

Permian

Carboniferous

Devonian

Silurian

27

VOLCANOES AND GEYSERS

Volcanoes and geysers provide glimpses of the Earth during its early evolution. When the crust was first forming, its surface was probably covered with many active volcanoes.

Volcanoes erupt when pressure increases in parts of the upper mantle. Then the molten rock, or magma, rises through the crust. Lava, which is magma when it reaches the surface, spews out of the volcano. Pieces of rock, ashes, and gases are sent flying. The rock and ash build up around the opening of the volcano, forming a cone. In time the cone may grow into a large mountain.

Geysers are often found in areas surrounding the volcanoes. Underground water is heated by magma deep in the Earth. Spurts of boiling water and steam shoot out of the geyser. Some geysers erupt on a regular basis. One of them, Old Faithful in Yellowstone National Park, erupts almost once an hour.

Although geysers have little effect on the Earth's surface, volcanoes are still a major influence today. For one thing, they continue to change the landscape and produce igneous rocks. Perhaps even more important is their effect on people. Volcanoes continue to cause massive destruction of property and can cause death or injury to people living nearby. In addition, volcanoes on the ocean floor may cause huge surges of water to flood coastal areas. The **tsunami**, also known as a tidal wave, often comes without warning, drowning any victims in its path. Major eruptions put dust high into the atmosphere. This blocks the sun and can affect climate for several years. Gases were released into the atmosphere following the eruption of Mt. Pinatubo in the Philippines in 1991. As a result, the average surface temperature around the world was lowered by 1° F.

1., 2., 3. Yellowstone National Park, in Wyoming, is a popular tourist attraction. A geyser shoots a plume of water and steam into the sky (Figure 1), while hot springs collect steaming pools of water (Figure 3). The terraces found at Mammoth Hot Springs in Yellowstone are the result of limestone deposits (Figure 2).

1

2

3

4

5

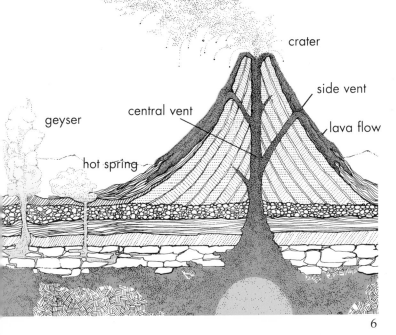

6

4. A large crater, called a caldera, has formed at the top of a volcano in Hawaii.

5. Lava hardened as it flowed from another volcano in Hawaii.

6. The internal structure of a volcano shows layers of hardened ash and lava that built up from a series of eruptions. Lava flows up the central vent and out through the volcano's crater. Lava may also flow out through side vents.

GLACIERS

Glaciers form when more snow falls in the winter than melts during the summer. When this happens, the snow settles in layers, getting deeper and deeper over the years. The bottom layers are compacted and recrystallize into ice. In the end, a glacier—a thick, moving mass of ice—is formed. Like a river, a glacier moves downhill. But that trip may be very slow. Once a glacier forms, it may not melt for hundreds, or even thousands, of years.

Glaciers move when their bottom layer slips along the ground. They also move because the mass of ice builds up so much pressure that the solid block of ice flows almost like a liquid. Although this movement may be fast—as much as several feet per day, it is usually very slow.

Small glaciers can be found in mountainous regions. Continental glaciers, also called ice caps, cover much larger areas. The continental glacier found in Antarctica covers an area almost one and a half times the size of the United States. Today ice covers the two poles and most of Greenland. But during the Pleistocene Epoch, much larger areas of the Earth were covered. This time period, from 2 million to 20 thousand years ago is commonly called the Ice Age. During this time, the Earth cooled and warmed several times, and glaciers formed, moved, and melted.

When glaciers move, they leave their imprint on the landscape. As glaciers move down mountains, they tear off pieces of rock as they go. Carving out the rock leaves broad, U-shaped valleys in the glacier's path. Rocks and other debris are left behind when the glacier melts, forming a ridge called a **moraine**.

While the wide valleys and steep peaks of such impressive mountains as the Rockies and the Alps were carved by relatively small glaciers, continental glaciers change the landscape in other ways. These glaciers are so massive that they carry away much in their path. As a result they generally flatten and soften the landscape.

1

2

1. At 5 1/2 miles above sea level, Mount Everest in the Himalaya Mountains has been known as the tallest mountain in the world. Its very steep sides, which are the ultimate challenge for mountain climbers, were formed by the action of glaciers. The Everest group of mountains is shown here.
2. The Khumbu Glacier continues to change the landscape of the Everest group.

moraine

3. Wide valleys are formed as a glacier moves between mountain peaks. Rocks picked up by the glacier are deposited in ridges called moraines.

3

RIVERS AND LAKES

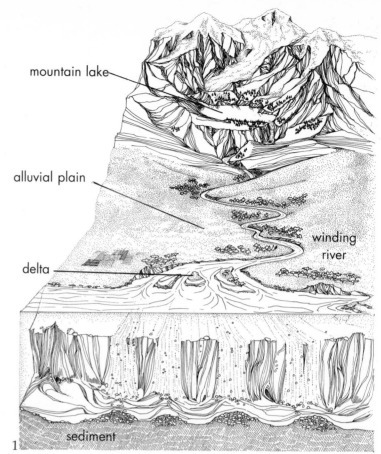

Like glaciers, rivers have a major impact on the landscape. Rivers carve deep, V-shaped valleys as they run downhill, eroding, or carrying away, rocks in their path. Then rocks picked up by the flowing water continue to wear away the bedrock in the river's path:

Rivers may start as trickles of water that wind their way downhill. At some point several brooks may join together to form a small stream, which follows the slope of the land. Several such streams may join together to form a river. And the young river follows a fairly direct path downhill, cutting into the rock as it flows. In time this young river erodes the rock bed to form a narrow, V-shaped valley. Then, as the river matures, it no longer cuts deep into the rock but begins to erode the sides of the riverbed, carving out a wider river valley with sloping sides. Once the valley has widened into a large floodplain, the river may begin to meander, or make wide curves, as it continues to flow downhill. As the river changes its course, it drops off some of the **sediment**, or debris, that it has carried, building up some areas of land, while eroding others. A river deposits the rest of its sediment when it reaches the ocean. At the mouth of the river the sediment may form a **delta** and separate the water into several channels. Or the river may empty into a broad estuary.

Rivers and streams may also deposit their water into lakes. Large lakes can form in craters of old, dead volcanoes or in the cavities left behind by glaciers. When a glacier remains in one spot and melts, it forms a large hole in the ground. Lakes forming in these depressions are called kettle lakes.

1. A mature river flows from a mountain lake, meandering through a broad floodplain, on its way to the ocean. Deltas form at the mouth of the river, with the remaining sediment washing out at the shore.
2. A kettle lake, over 3 miles deep, found in the Himalaya Mountains of Nepal.
3. The Iguaçu Falls in Brazil. A waterfall is the result of water running off a rocky ledge to an area below. Although young rivers may have waterfalls, mature rivers cut down into the rock, leveling the riverbed and eliminating the waterfall.
4. The rocky ledge of a waterfall is eroded at different rates, with soft rock eroding more quickly than harder rock.
5. The Grand Canyon in Arizona was formed by the action of the Colorado River.

3

hard rock

soft rock

4

5

FORMATION OF THE ATMOSPHERE

The Earth's atmosphere is made up mainly of two gases—78 percent nitrogen and 21 percent oxygen. But when the Earth first formed, several billion years ago, there was no atmosphere. As you have read, gases and any very light elements were swept away from the Earth by powerful solar winds. But in time, a relatively thin layer of gases formed around the Earth.

Where did the atmosphere come from? Scientists believe that the first atmosphere was made up of gases that had been dissolved in magma deep below the crust. As the crust formed, many volcanoes erupted, releasing these trapped gases along with the molten rock. And those gases were probably the same ones that are released from volcanoes active today. Although it is difficult and dangerous to collect samples from volcanoes, scientists have determined that most of the gas released is water vapor, with smaller amounts of carbon dioxide, nitrogen, sulfur compounds, chlorine, and hydrogen. So the atmosphere long ago would probably have been made up of those gases.

Oxygen was not one of the gases of the early atmosphere. It was added over time. Very simple forms of life appeared over 3 billion years ago. Some early types of living things were photosynthetic—like plants, they could make their own food from carbon dioxide and water. Along with food, they also gave off oxygen. After many years of oxygen being released from photosynthetic organisms, the level of oxygen built up within the atmosphere.

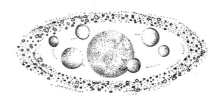

5 billion years ago the Earth was formed as part of the solar system.

4.6 billion years ago the Earth's crust formed.

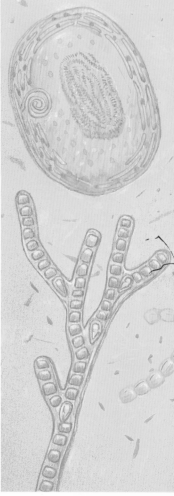

1. When the Earth first solidified, it had no atmosphere. As volcanoes erupted through the newly-formed crust, gases were released and collected around the Earth, making up the first atmosphere.

As the Earth cooled, simple life-forms sprang up in the ocean, and at some point, plantlike organisms evolved. These organisms took in energy from sunlight and gave off oxygen, releasing it into the newly forming atmosphere. In time the oxygen level of the atmosphere rose, and a very thin layer of ozone, an oxygen compound, formed high up in the atmosphere. This ozone layer protects all living things from the harmful effects of ultraviolet radiation given off by the sun.

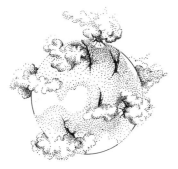

4.5 billion years ago the ocean and original landmasses formed.

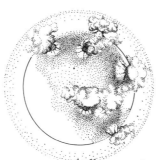

2.5 billion years ago the atmosphere was formed.

280 million years ago Pangea was formed.

2

2. The atmosphere as seen from a satellite.

3. These stromatolites, found in Shark Bay, Australia, are thick mats made up of bacteria, blue-green bacteria, and sand. Stromatolites may have contributed oxygen to the early atmosphere—fossils of similar mats have been dated at more than 3 billion years old.

WEATHER AND CLIMATE

Weather can change from day to day. But climate does not change so quickly. The climate of an area is its general weather or the condition of the atmosphere over a period of time.

Climate depends on temperature, moisture, air pressure, and wind. Temperature depends mainly on the area's location—its latitude and altitude. In general, the farther from the equator and the higher up from sea level, the cooler an area will be.

warmth is transferred through a large volume of water. Since a large volume of water has been warmed, it cools down very slowly in the evening.

The ocean also affects climate by adding moisture to the air. Water evaporates from the ocean. As water is heated, it **evaporates**, or changes from a liquid to a gas. This gas, called water vapor, is carried up into the atmosphere. Because the air temperature is cooler higher in the atmosphere, the water vapor is cooled. As it cools it condenses, or changes back into a liquid. Tiny water droplets then gather together, forming clouds. As the droplets grow in size, they may fall back to Earth as precipitation—rain, snow, or hail. This journey of

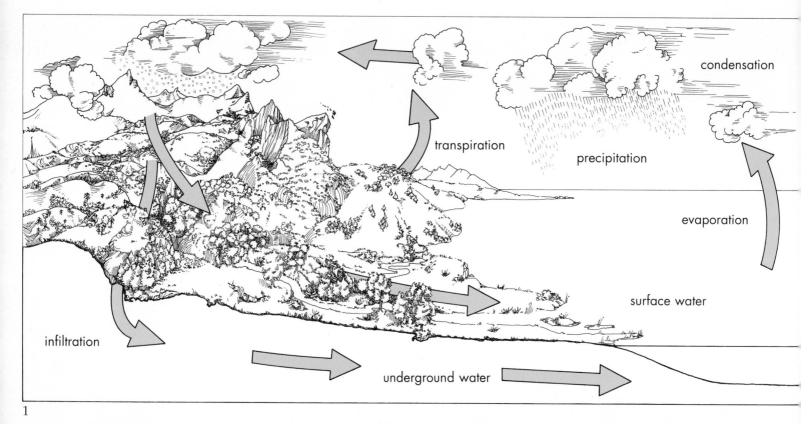

1

1. The water cycle. Water enters the atmosphere through evaporation or transpiration. Water falls back down to the surface as precipitation, supplying water to living things as well as to bodies of water and to underground water supplies.

Areas near the equator are warm because they receive the sun's rays directly, not at an angle. As you travel away from the equator, the sun's rays strike the Earth at a greater and greater slant. As a result, the same amount of energy that shines on a small area near the equator is spread out over a much larger area. Because this larger area receives less solar energy, it is cooler than areas near the equator.

When the sun shines on land, only the surface of the land is affected, and it quickly warms up. In the evening, the land quickly cools down again. But when the sun shines on water, the water temperature rises more slowly. This is because the

water, from the Earth's surface to the atmosphere and back, is called the **water cycle**. It has no beginning and no end. One important part of the water cycle is water given off by plants to the atmosphere. As plants pull water through their systems, they give off water through pores in their leaves. This process is called **transpiration**.

2. This photo of a cyclone taken from a weather satellite shows the typical swirling pattern.
3. Cumulus clouds are also known as fair-weather clouds.
4. Electrical discharges from clouds result in spectacular flashes of light known as lightning.

4

3

Pages 38-39:
In one area, over the course of millions of years, the climate can change greatly. During the three periods illustrated, the climate changes from glacial (1), to tropical (2), to desert (3).

A glacial climate

A tropical climate

A desert climate

R. MORIGGIA & M. PIATTO

DESERTS

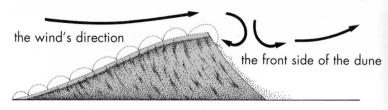

the wind's direction

the front side of the dune

a dune seen from the side

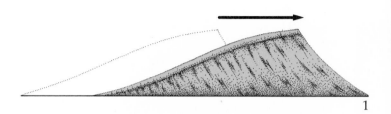

1

If you think of a desert as a place that is hot and dry, and where nothing lives, you would be partly right. Some deserts are like that. But there are also very cold deserts near the North and South Poles, where all moisture freezes and is unavailable to plants. Many deserts that are hot while they are warmed by the sun, quickly cool down in the evening. In addition, many deserts and arid, or dry, regions do support living things. In these dry areas, however, there are fewer species, or types of living things, than in milder climates. So deserts are not necessarily hot and void of living things, but they are dry. Deserts are usually defined as areas that receive less than 10 inches of rainfall per year.

One of the most visible factors in shaping a desert is the wind. In most other types of landscapes, plants hold the soil in place. But the sparse plant life of the desert allows the wind to pick up desert soil and sand. Just as glaciers and rivers can erode, or wear away the landscape, wind can also cut into rock and change the shape of the land.

Wind **erosion** can lead to a feature known as desert pavement, a level layer of small rocks and pebbles, much like a vast parking lot, paving the desert floor. Once the sand and soil are picked up, only the heavier particles, such as the pebbles and rocks, are left behind, forming the desert pavement.

Desert pavement protects soil below from further erosion. If it is broken, that soil is once again exposed to rapid wind erosion. However, desert pavement takes a long time to form. That is why dirt bikes and dune buggies, which tear up the pavement, have such disastrous impact.

The wind also moves sand and soil into dunes, or small hills. If there is enough plant life, it may grow on the dunes and stabilize them. Without plants, the dunes are built up and swept away, only to reform some distance away.

1. In general, dunes continually move in the direction of the wind. Wind blows over the crest of the dune and deposits sand at the base of the dune on its wind-sheltered side. More and more grains are picked up and deposited, resulting in the entire dune moving forward.
2. Crescent-shaped dunes form when a prevailing wind blows from a single direction.

3. Rocks emerge from the sand in the Sahara Desert in Niger.
4. An oasis, or relatively lush area, sometimes forms within deserts. This one is in found in Algeria.
5. Soil in this arid region of Algeria has cracked from lack of rain.

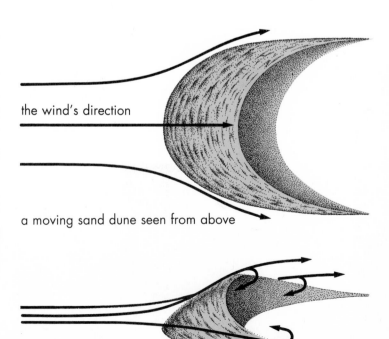

the wind's direction

a moving sand dune seen from above

a moving sand dune seen from the side

2

3

4

5

GLOSSARY

anticline: a fold in the strata in the shape of an arch.

atoll: a type of coral reef often found in tropical waters, which is usually in the shape of a curve or ring, surrounding a lagoon.

core: the innermost layer of the Earth, made up of a molten outer core and a solid inner core.

crust: the outer rocky layer of the Earth.

cyclone: a low-pressure center that causes great masses of air to whirl and rotate.

delta: an accumulation of sediment forming at the mouth of a river where it enters a lake or the ocean.

dome mountain: a mountain that forms where geologic forces lift the Earth's crust into a bulge or dome.

erosion: the removal of rock or other materials by the action of water, wind, or ice.

evaporation: an action by which matter changes from the liquid state to the gaseous state.

fault: a crack in a series of strata, or layers of the Earth.

fault-block mountain: a mountain that forms when huge blocks of the Earth's crust push up along a fault.

fold: a change in the shape of a rocky area caused by geological forces.

folded mountain: a mountain that forms when two plates meet and their edges crumble.

geyser: a hot spring that shoots up hot water and steam.

graben: a valley formed by the downward movement of a fault block.

granite: a coarse and large-grained rock that formed as magma cooled slowly deep in the Earth. It is the main igneous rock of the continental crust.

igneous rock: rock formed from magma.

inner core: the ball-shaped innermost structure of the Earth, consisting of solid iron and nickel.

limestone: a sedimentary rock composed entirely or almost entirely of calcium carbonate.

magma: molten underground rock.

mantle: the layer of rock below the Earth's crust.

metamorphic rock: rock that is changed from igneous or sedimentary rock by being under great heat and pressure.

mid-ocean ridge: an area of the ocean floor where magma pushes from below, reaching the surface as the ocean floor spreads.

mineral: a crystalline material that is made up of one or more elements that occur naturally in the Earth.

mollusk: a soft-bodied animal usually covered by a shell. Examples include snails and clams.

moraine: deposits of rock that have been left by glaciers at their edges or where they end.

nebular hypothesis: hypothesis that the solar system formed from the condensation of a huge, swirling mass of gas and dust.

outer core: a layer of the Earth that begins about 1,800 miles beneath the surface and is about 1,400 miles thick. It is made up of melted iron and nickel.

plates: large sections of the crust that are moved by molten material within the mantle.

precipitation: moisture which falls from the atmosphere to the surface of the Earth in the form of rain, snow, or hail.

protoplanet: a mass of matter before it condensed to form a planet.

rift: a long, deep crack. One example is the rift that runs up and down the highest part of mid-ocean ridges.

rock cycle: the process by which rocks change from one type to another: igneous, metamorphic, or sedimentary.

sediment: eroded rock material deposited on the Earth's surface by water, wind, and ice.

sedimentary rock: rock formed from sediment by cementation and/or pressure.

silicon: a widely found nonmetallic element.

spring: a source of underground water that gushes out onto the Earth's surface.

strata: broad layers of sedimentary rock.

syncline: a U-shaped fold in the strata.

transpiration: an action or process by which plants give off water into the atmosphere.

tsunami: a Japanese word meaning a large and destructive wave caused by an oceanic earthquake, also called a tidal wave.

volcanic mountain: mountain that forms as a result of volcanoes.

water cycle: the movement of water between and through the solid Earth and the atmosphere.

water vapor: water in a gaseous state.

FURTHER READING

Aylesworth, Thomas G. *Moving Continents: Our Changing Earth.* Enslow, 1990

Catherall, Ed. *Exploring Soil and Rocks.* Raintree Steck-Vaughn, 1990

Cork, B. and Bramwell, M. *Rocks and Fossils.* EDC, 1983

Dineen, Jacqueline. *Volcanoes.* Watts, 1991

Fisher, David E. *The Origin and Evolution of Our Own Particular Universe.* Macmillan, 1988

Grady, Sean M. *Plate Tectonics: Earth's Shifting Crust.* Lucent Books, 1991

Knapp, Brian. *Earthquake.* Raintree Steck-Vaughn, 1990

Simon, Seymour. *Mountains.* Morrow, 1994

INDEX

44

EVOLUTION OF THE MONERAN, PROTIST, PLANT, AND FUNGI KINGDOMS

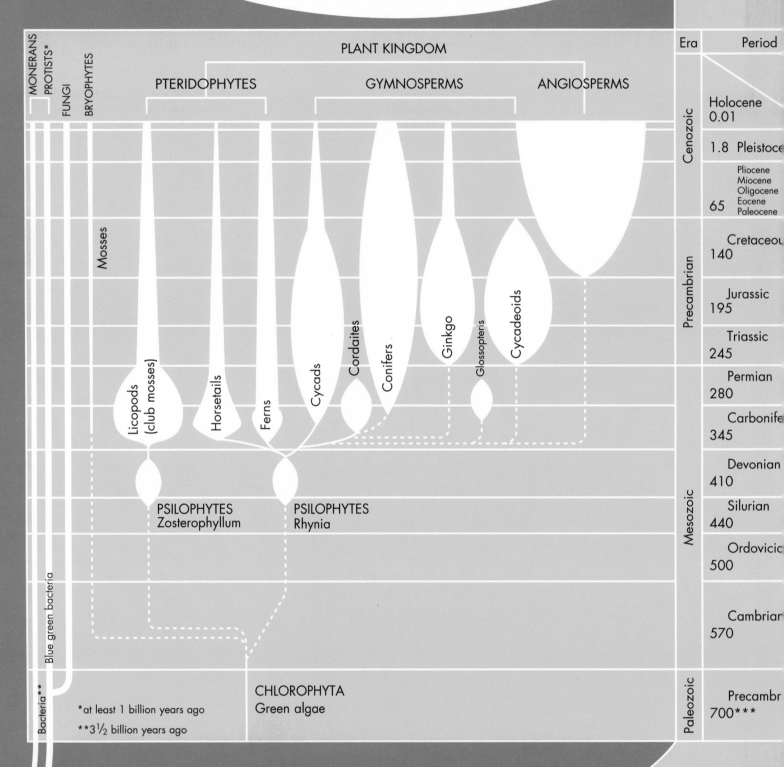

MONERANS
PROTISTS*
FUNGI
BRYOPHYTES

PLANT KINGDOM

PTERIDOPHYTES   GYMNOSPERMS   ANGIOSPERMS

Mosses

Licopods (club mosses)

Horsetails

Ferns

Cycads

Cordaites

Conifers

Ginkgo

Glossopteris

Cycadeoids

PSILOPHYTES
Zosterophyllum

PSILOPHYTES
Rhynia

Blue green bacteria

Bacteria**

CHLOROPHYTA
Green algae

*at least 1 billion years ago

**3 1/2 billion years ago

Era	Period
Cenozoic	Holocene 0.01
	1.8 Pleistoce
	Pliocene Miocene Oligocene Eocene Paleocene
	65
Precambrian	Cretaceou 140
	Jurassic 195
	Triassic 245
Mesozoic	Permian 280
	Carbonife 345
	Devonian 410
	Silurian 440
	Ordovici 500
	Cambriar 570
Paleozoic	Precambr 700***

EVOLUTION OF THE PROTIST AND ANIMAL KINGDOMS

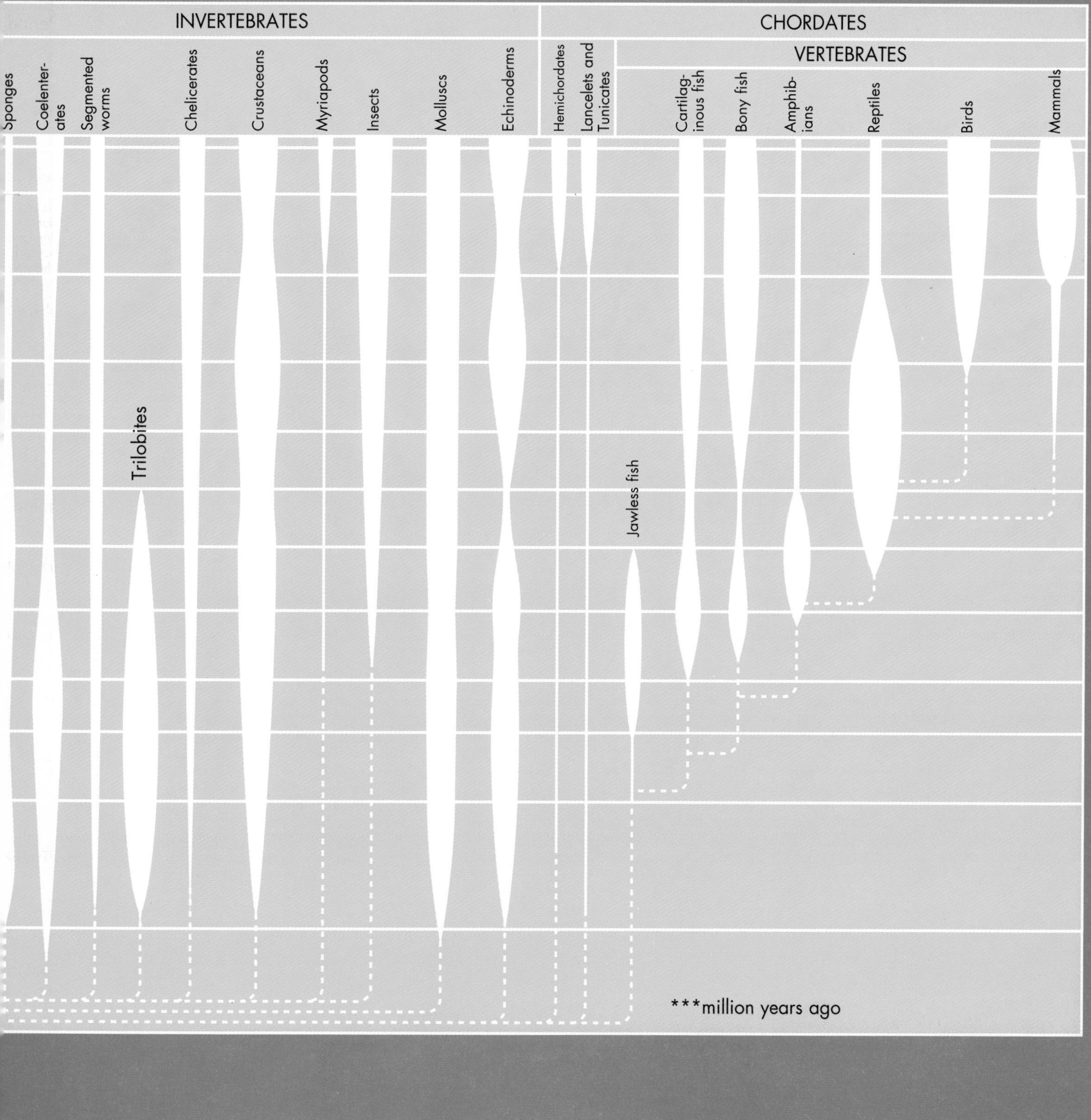

INVERTEBRATES

CHORDATES

VERTEBRATES

Sponges

Coelenterates

Segmented worms

Chelicerates

Crustaceans

Myriapods

Insects

Molluscs

Echinoderms

Hemichordates

Lancelets and Tunicates

Cartilaginous fish

Bony fish

Amphibians

Reptiles

Birds

Mammals

Trilobites

Jawless fish

***million years ago

YA

550
ALE   Alessandrello, Anna

           The earth